SpringerBriefs in Applied Sciences and Technology

More information about this series at http://www.springer.com/series/8884

Jaap Schijve

Biaxial Fatigue of Metals

The Present Understanding

 Springer

Jaap Schijve
Delft University of Technology
Delft
The Netherlands

ISSN 2191-530X   ISSN 2191-5318   (electronic)
SpringerBriefs in Applied Sciences and Technology
ISBN 978-3-319-23605-6   ISBN 978-3-319-23606-3   (eBook)
DOI 10.1007/978-3-319-23606-3

Library of Congress Control Number: 2015948570

Springer Cham Heidelberg New York Dordrecht London

Springer International Publishing AG Switzerland is part of Springer Science+Business Media
(www.springer.com)

Contents

Abstract

Problems of fatigue under multiaxial fatigue loads have been addressed in a very large number of research publications. The present paper is primarily a survey paper of biaxial fatigue under constant amplitude loading on metal specimens. It starts with the physical understanding of the fatigue phenomenon under biaxial fatigue loads. Various types of proportional and non-proportional biaxial fatigue loads and biaxial stress distributions in a material are specified. Attention is paid to the fatigue limit, crack nucleation, initial micro-crack growth and subsequent macro-crack in different modes of crack growth. The interference between the upper and lower surfaces of a fatigue crack is discussed. Possibilities for predictions of biaxial fatigue properties are analysed with reference to the similarity concept. The significance of the present understanding for structural design problems is considered. The paper is completed with a summary of major observations.

Keywords Multiaxial fatigue load sequences · Physical fatigue phenomenon · Biaxial fatigue experiments · Crack nucleation and growth · Predictions on multiaxial fatigue

Biaxial Fatigue of Metals

1 Introduction

The development of technical knowledge about fatigue of materials and structures started in the 19th century. Several noteworthy events occurred including serious accidents with dramatic fatalities and failures of large structures. Impressive early research was carried out by Wöhler in the middle of the 19th century. He investigated fatigue failures of railway axles which occurred at a severe stress concentration and with fretting corrosion, see Fig. 1a. Wöhler introduced a modification of the clamped connection between the wheel and the axle, see Fig. 1b[1]. The severe stress concentration and fretting corrosion were eliminated. It is an illustrative example of *"designing against fatigue"*. Extensive research on fatigue of structures and materials has been carried out in the 20th century which revealed a large variety of relevant aspects of fatigue properties. At the same time fundamental research was focussed on the fatigue phenomenon in metallic materials. By now crack nucleation, micro crack growth and crack propagation until failure are reasonably well defined and qualitative understood in physical terms. Unfortunately, the understanding is sufficient to arrive at the conclusion that quantitative predictions can not be achieved [2]. It has led to a variety of service-simulation fatigue tests.

Fatigue tests for a long time were carried out with uniaxial fatigue loads. However, it was recognized that multiaxial stresses can occur in full-scale structures. A well known case is a pressurized container used in various production equipment. The aircraft fuselage is another classical example. The circumferential hoop stress and the longitudinal tension stress are two mutually perpendicular components of a biaxial stress condition. Actually a uniaxial stress distribution

---

[1]Figure copied from [1].

© The Author(s) 2016
J. Schijve, *Biaxial Fatigue of Metals*, SpringerBriefs in Applied
Sciences and Technology, DOI 10.1007/978-3-319-23606-3_1

(a)   **(b)**

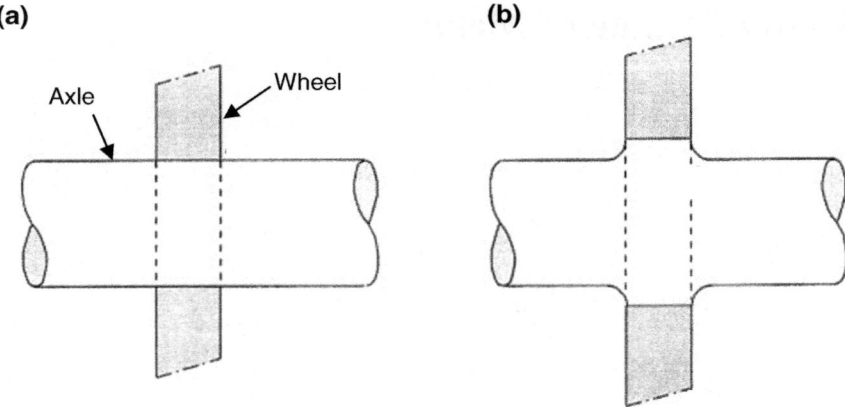

Fig. 1 Designing against fatigue of a railway axles in the 19th century by Wöhler. **a** Poor design.
b Improved design

rarely occurs in engineering structures. It can occur in slender beams of a structure,
but the joints of a beam are meeting with a biaxial stress condition. So we know for
sure that biaxial stress distributions do occur in a structure. It then would be strange
if we would not be interested in this physical phenomenon occurring in structures.

The major intention of this paper is to arrive at the present understanding of
fatigue under constant-amplitude (CA) biaxial loading. The term multiaxial fatigue
is avoided because it includes triaxial loading. Triaxial loading is rarely addressed
in the literature. Moreover at the surface of a material the stress condition is biaxial.
Furthermore, various fatigue problems under uniaxial loading have not been solved
with rational arguments. It then can not be expected that they can be solved for
biaxial loading. As an example this applies to fatigue under variable-amplitude
loading. The consequence is that the present survey of biaxial fatigue load problems
is restricted to CA biaxial fatigue. In spite of these restrictions, the number of papers
on multiaxial fatigue is still very large. It was noted by Marquis as a result of an
internet search that in 2003 more than 1500 papers had been published on multiaxial fatigue and fracture [3]. Since then the number of publications did still steadily
increase. The very high number of research programs is associated with the large
variety of possible biaxial load conditions, the various types of specimens, and
different specimen materials adopted. In contrast to the abundant publications it is
remarkable that the interest of engineers in the design office of the industry for
multiaxial fatigue is practically absent. In order to explain this gap the physical and
engineering aspects of multiaxial fatigue are discussed.

The paper starts with a discussion of the fatigue phenomenon under uniaxial
loading and under biaxial loading. Attention is paid to fatigue crack nucleation,
micro crack growth and the macro crack propagation. A survey is presented of the
variety of biaxial load conditions applied in research investigations and the corresponding biaxial stress conditions obtained in specimens. It includes tension,

bending, torsion and shear loads and different crack growth modes. A survey is presented of various types of specimens adopted in biaxial fatigue investigations. Problems about predictions of biaxial fatigue properties are discussed with physical understanding as a leading theme. The engineering perception of dealing with biaxial fatigue properties is considered. The paper is concluded with a summary of major observations.

2 Physical Aspects of the Fatigue Phenomenon Under Uniaxial and Biaxial Loading

2.1 The Fatigue Phenomenon Under Uniaxial Loading

The fatigue life of a specimen under cyclic loading starts with an initial period in which a small micro crack is generated. It is followed by crack growth, initially still as a micro crack, and afterwards as a macro-crack until failure, see the diagram in Fig. 2 published by Shamsaei and Fatemi [4]. It includes some indicative dimensions of crack growth ranges. It is generally accepted that dislocation mechanisms are involved in crack nucleation and crack extension mechanism because of cyclic slip in a crystalline material. The physical mechanism about how cyclic slip is causing the micro crack initiation and subsequent micro crack growth is still not exactly understood. It is a decohesion phenomenon. The initiation of a fatigue crack occurs preferably at the material surface because the restraint on cyclic slip is less effective at a free material surface.

General experience has shown that the fatigue mechanism depends on the type of material and its particular structural characteristics which are the crystal lattice, potential slip systems, cross slip, elastic anisotropy, texture, inclusions, impurities, grain size, grain shapes, pearlite bands in mild steel, etc. It may be noted here that

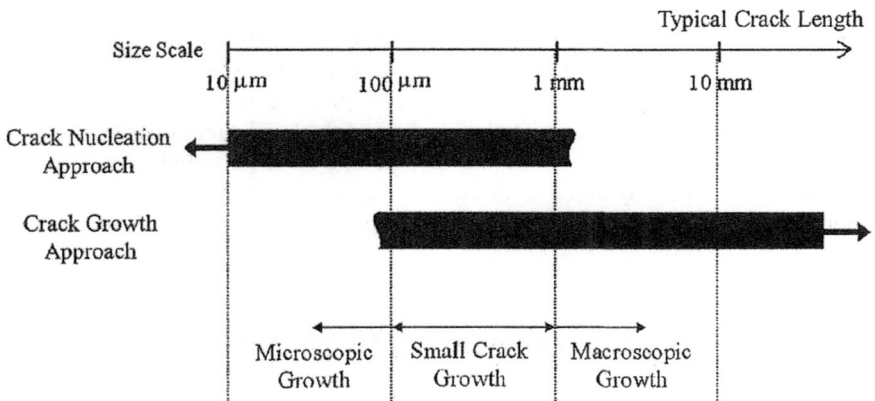

Fig. 2 Fatigue crack growth in three periods defined by Shamsaei and Fatemi [4]

Fig. 3 Top view of crack
with crack front passing
through many grains [1]

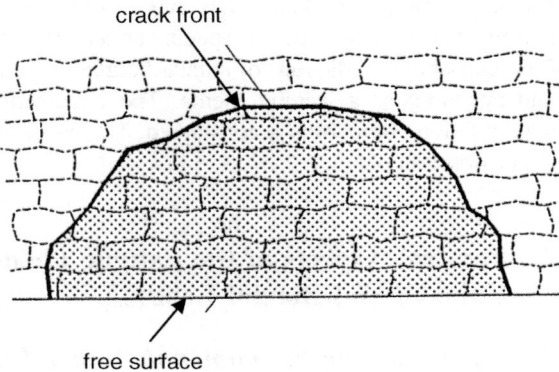

the material of specimens in many papers on multiaxial fatigue is considered to be homogeneous and isotropic, and to have well defined properties, a drastic simplification to be kept in mind.

The growth of the very first micro crack will be retarded when the crack tip is approaching the first grain boundary. Subsequent growth into adjacent grains along the material surface and away from the material surface is initially a matter of overcoming such barriers. The very first micro crack growth is a surface phenomenon which still occurs in a somewhat irregular way, also because it may start from subsurface inclusions or impurities. However, after some further crack growth the behaviour becomes essentially different. The micro crack then has penetrated into more grains and the crack is simultaneously growing with a coherent crack front in a substantial number of grains, see the schematic picture in Fig. 3. Crack growth now is depending on the crack growth resistance of the material which is a kind of a bulk property of the material. It is no longer depending on the free surface of the material. A more extensive description is presented in [5].

It should be understood that the conversion from the initial surface phenomenon to a micro crack with a realistic crack front is a transient process. The conversion does not occur at a sharply defined moment in the fatigue life, or at a specific size of the still small crack. As suggested in Fig. 2 it may occur in the range of 0.1–1.0 mm. Predictions on the crack growth rate in the three crack length regions offer different problems. In the crack nucleation period the local stress distribution at the material surface is important. In the last period of macroscopic crack growth concepts introduced by fracture mechanics can be considered. However, in the intermediate period predictions are problematic.

2.2 *Different Modes of Fatigue Crack Growth*

Under uniaxial fatigue loads the fatigue crack is generally growing in a transgranular way with a tendency to grow in a direction perpendicular to the main

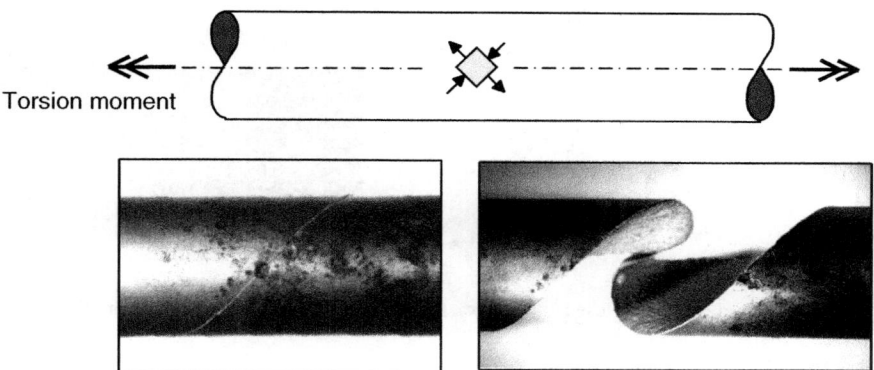

Fig. 4 Fatigue failure of a torsion spring of a motor-car. Crack started at surface pit [1]

principal stress. This is also true for a cylindrical bar loaded by a cyclic torsion load, see Fig. 4. The crack opening occurs in the direction of the tensile stress. It is referred to as a Mode I case of crack opening. Actually, three different modes can be defined for crack opening, see Mode I, Mode II and Mode III in Fig. 5. They are characterized by different directions of the crack opening. The more important ones are Mode I and Mode III.

Interesting observations were recently published by Tanaka [6]. He carried out fatigue tests on specimens with a blunt notch and a sharp notch, see Fig. 6. The specimens were tested under cyclic torsion with and without static tension. Small cracks were initiated as Mode I cracks, actually the same behaviour as illustrated in Fig. 4. However in the root of the notch many cracks were initiated simultaneously in two oblique directions. It leads to the so-called factory roof topography, see Fig. 6. It is important to realize that such a fatigue fracture surface is far from being a flat surface.

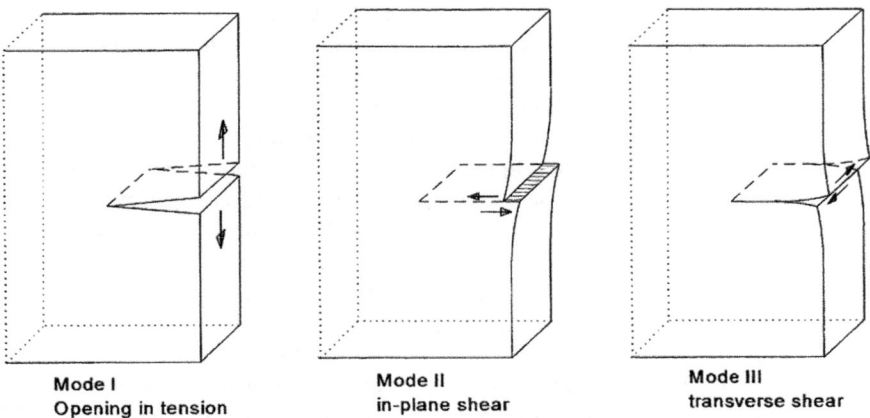

Mode I	Mode II	Mode III
Opening in tension	in-plane shear	transverse shear

Fig. 5 Three different crack opening modes

(a)   **(b)**

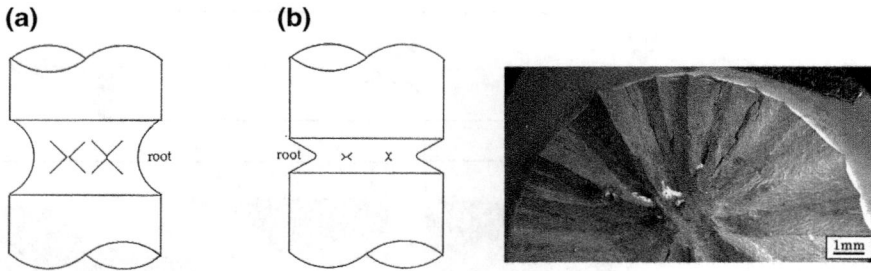

Fig. 6 Notched specimens used by Tanaka for fractographic observations [6]. **a** Blunt notch.
b Sharp notch

2.3 The Fatigue Phenomenon Under Biaxial Load Cycles

The discussion may well start from the proposition that crack nucleation and crack
growth under biaxial load conditions are associated with cyclic slip. Although the
biaxiality will affect the phenomenon, the initiation will still occur as a surface
phenomenon followed by micro crack growth and macro crack propagation as
discussed in Sect. 2.1. However, the biaxial stresses will affect the cyclic slip and
the conversion into crack nucleation and crack extension. A significant effect may
well be expected. The problem how cyclic plasticity is physically leading to the
very first crack nucleus and subsequent crack extension is difficult, and in fact
beyond the present understanding. In the literature cyclic plasticity is discussed in
the evaluation of so-called critical plane models. It is realized that cyclic strain
hardening can occur in persistent slip zones but plastic shake down can also occur.
It depends on the material. Some physical speculation about biaxial fatigue phe-
nomena is possible, but realistic predictions can not be made.

The macro crack growth period offers another problem. It is associated with the
question whether crack growth occurs with the crack being open or closed.
Interesting experiments were carried out by Tschegg [7] with a specimen shown in
Fig. 7, a specimen with a sharp circumferential notch. Under cyclic torsion a
Mode III crack was growing inwards. The second biaxial load was a tension load or
a compression load perpendicular to the fatigue fracture surface. A tension load
opens the Mode III crack. However, the compression load will close the crack. In
the latter case the torsion load will cause a cyclic sliding contact between the upper
and lower fatigue crack surfaces. As a consequence the crack growth rate was
considerably reduced. Under various biaxial load experiments such interferences
between the two surfaces of the fatigue crack can occur. The sliding contact
mechanism was also observed by Yu and Abel [8]. They are using the term "crack
surface interference" (CSI). The interference can also occur if a crack is not closed.
It can occur due to a non-flat crack surface topology. It was recently observed by De
Freitas et al. [9, 10] in rotating bending tests carried out with and without a steady
torsion load. The torsion load reduced the crack growth rate. With SEM

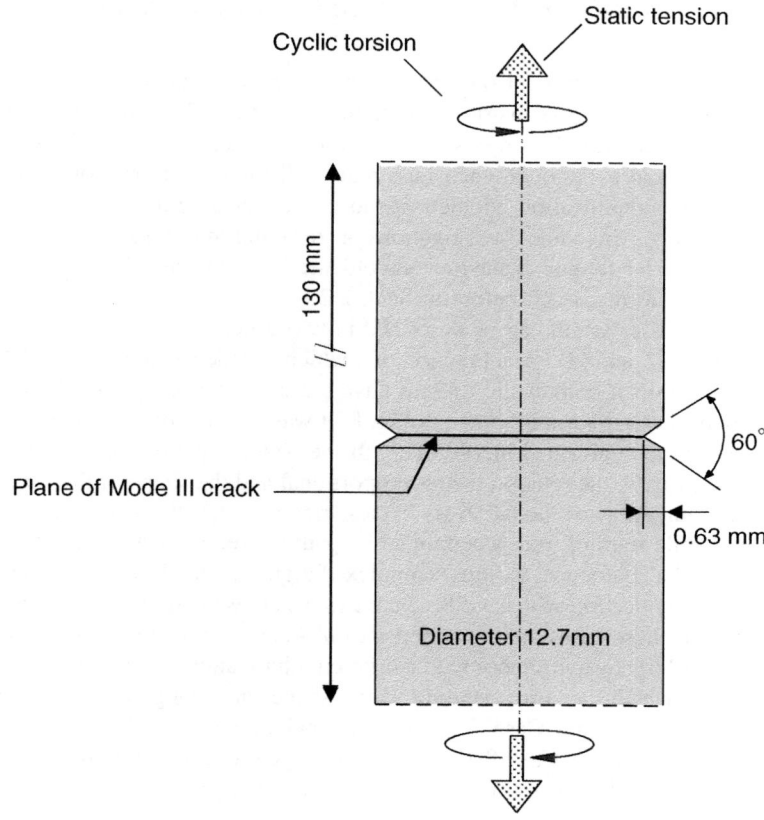

Fig. 7 Specimen loaded under cyclic torsion and a static tension or compression load used by Tschegg [7]

observations it was shown that the torsion load was causing a factory roof fatigue fracture surface. This undulated surface caused crack surface interference with the reduced crack growth rate as a result.

3   Biaxial Fatigue Research Programs

An extensive introduction to multiaxial fatigue was presented in the book published by Socie and Marquis [11]. An instructive introduction was presented by Fatemi [12].

3.1 Two Methods to Describe Biaxial Load Conditions

The biaxial stress condition introduced in a specimen or a structure is the result of a superposition of two stress distributions induced by two different loads. For simplicity of the discussion the stress systems of the two biaxial loads are indicated by the symbols σ and τ, although each load may well induce a more complex stress distribution. A simplification adopted in most research programs is to consider a zero mean stress. In some investigations a zero minimum load was adopted. A different biaxial fatigue behaviour should then be expected because of crack opening effects as discussed before in Sect. 2.2.

A biaxial stress system can be described in two different ways: (1) σ and τ as a function of time t, and (2) τ as a function of σ which is called a stress path. The most simple case occurs if both cyclic stresses have the same frequency and phase angle. This proportional variation is shown in Fig. 8. It was adopted by Gough et al. in the early investigation published in 1935 [13]. If the phase angle of the two stresses is different, see Fig. 9, the relation is non-proportional and the shape of the τ/σ-path is elliptical. With the two biaxial stress cycles, $\tau(t)$ and $\sigma(t)$, the stress path can be described, while starting from a certain stress path the required stress cycles can be derived. In the literature various complex fatigue paths have been adopted. A noteworthy collection with 13 different cases was presented by Yuuki et al. [14], see Fig. 10. If these cases are converted into τ/σ-diagrams it turns out that several cases show different wave shapes with a different phase angle. This applies to Cases 12 and 13 for which the corresponding waves shapes are triangular and sinusoidal, respectively. Case 8 and Case 9 are of principal interest, see Fig. 11. The τ/σ-diagram in Fig. 11a for Case 8 shows that a τ–cycle occurs when σ is a tension stress. However, in Fig. 11b for Case 9 the τ-cycle occurs when σ is a compression

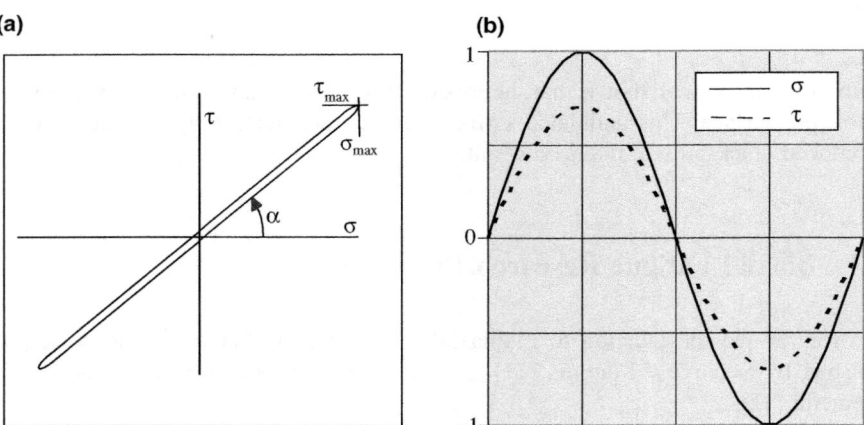

Fig. 8 Proportional loads presented in two different formats. **a** Proportional loading. **b** $\sigma(t)$ and $\tau(t)$

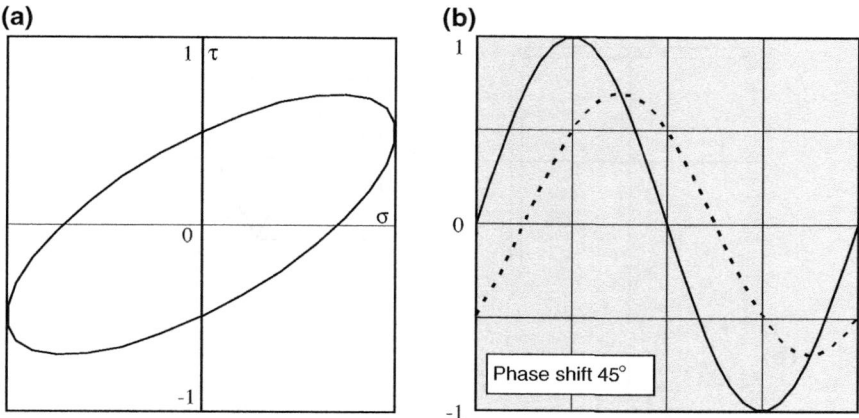

Fig. 9 Non-proportional loads presented in two different formats. **a** Non-proportional loading.
b $\sigma(t)$ and $\tau(t)$

stress. Physical consequences of the difference between the two cases were already discussed in Sect. 2.2 with a reference to research of Tschegg [7].

Complex conditions occur if two biaxial loads have different frequencies. It is illustrated in Fig. 12a with two sine wave loads. They are transferred into a τ/σ-path

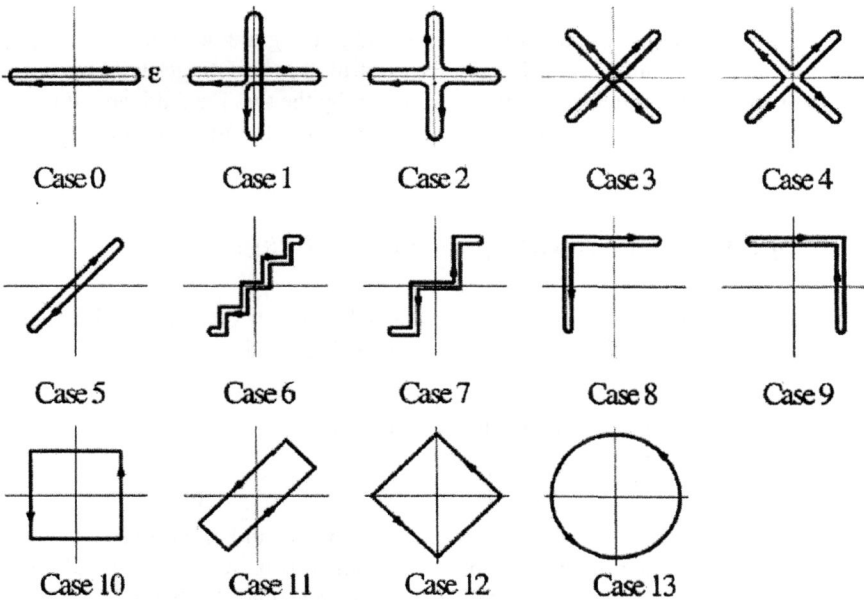

Fig. 10 Biaxial proportional and non-proportional paths presented by Yuuki et al. [14]

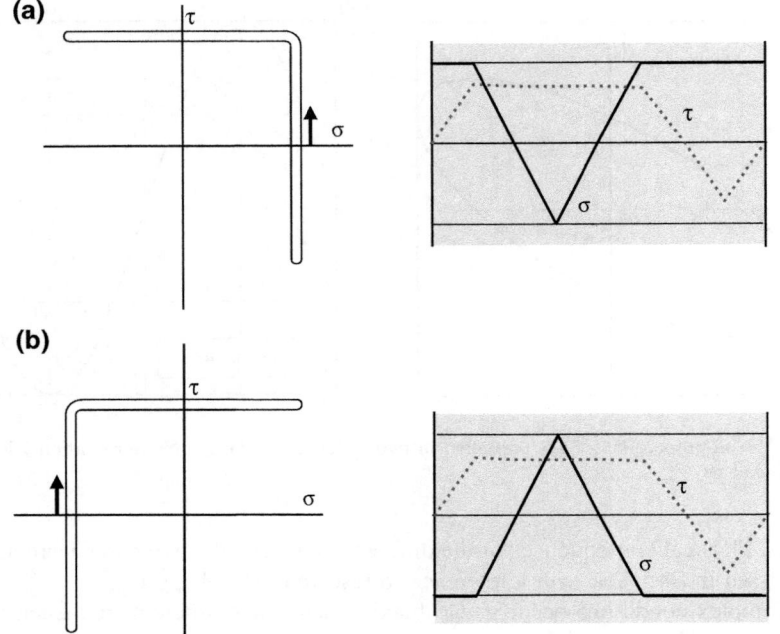

Fig. 11 Load cases 8 and 9 of Fig. 10 with the corresponding $\sigma(t)$ and $\tau(t)$

in Fig. 12b for a period of three times the period of the graph in Fig. 12a. It appears to be a chaotic picture for considering the contribution to fatigue damage, but the fatigue path is still contained within the limits set by the two biaxial loads. It may be said here in more popular terms, "the τ/σ-path is what the material feels".

3.2 Specimens for Research on Biaxial Fatigue

The variety of specimens used in experimental research on multiaxial fatigue is large. Four different types of specimens will be discussed.

- Solid specimens, unnotched and notched
- Tubular specimens, also unnotched and notched
- Cruciform specimens
- Sheet specimens for fatigue crack growth

The development of an experimental test set-up is a kind of a design problem. It includes the shape and dimensions of the specimen to be tested, but in addition a fatigue machine to apply the biaxial loads. Fatigue machines for biaxial fatigue tests are commercially available, but in many investigation a home made design is

(a)

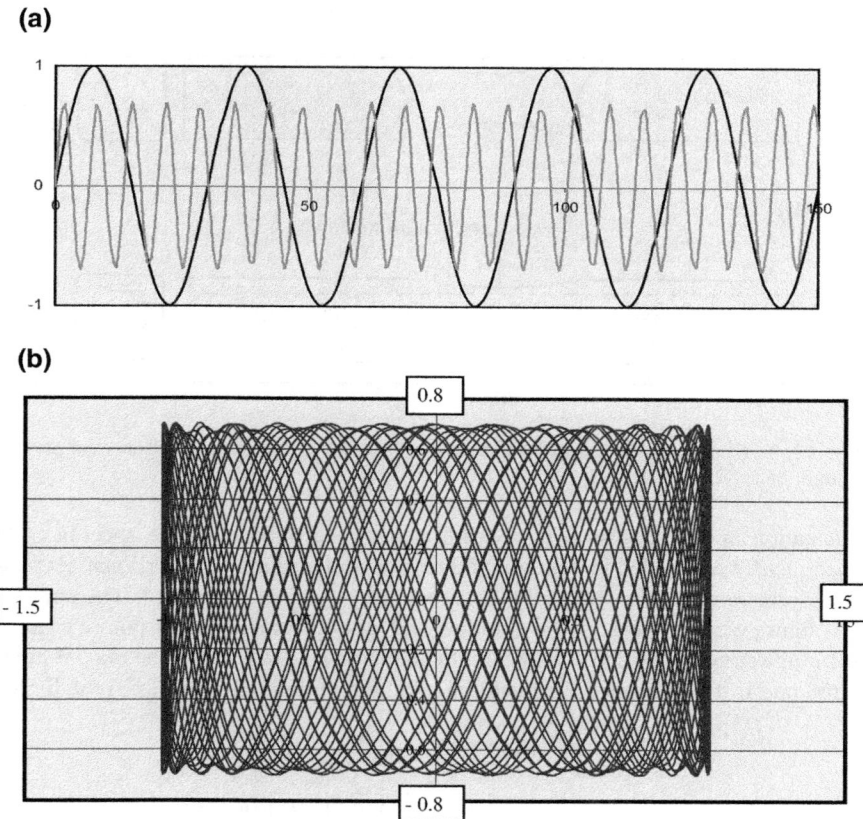

(b)

Fig. 12 Biaxial loads with a different frequency. **a** Two sine waves with a different frequency. The frequency ratio is 4.46. **b** The corresponding plot of σ/τ

developed to carry out biaxial tests with non-proportional biaxial loads. Since a fatigue machine is used in many tests it requires a robust design.

The production of specimens and the performance of the fatigue tests should be carefully planned in a similar way adopted for uniaxial fatigue tests. Clamping of a specimen is a generally recognized problem because failure in the clamping area must be avoided. The significant area of the specimen is the location where crack nucleation and crack growth will occur. Some typical specimens are discussed below with some comments on the purpose of the fatigue tests.

3.2.1 Solid and Tubular Specimens

In the early investigation of Gough and Pollard [13] two types of unnotched specimens were used, see Fig. 13. In the tubular specimen the visibility of crack

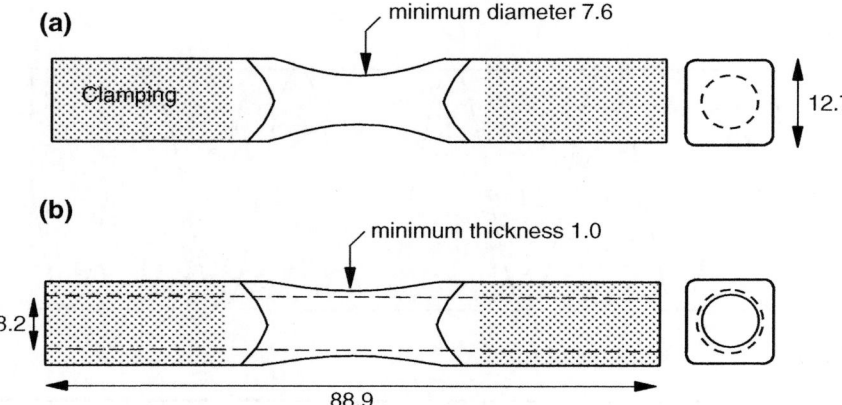

Fig. 13 Two specimens tested by Gough and Pollard [13]. Dimensions in figures are presented in mm

nucleation and propagation is concentrated at the outside of the specimen. The specimens were loaded in bending and torsion with different proportional τ/σ ratios. The corresponding fatigue limits obtained are presented in Fig. 14. The curves in this figure were proposed to be elliptical and labelled as a quadrant failure criterion. If τ/σ in a structure was inside the elliptical quadrant is was below the biaxial fatigue limit, and if it was outside the quadrant a fatigue failure might occur. Actually for a

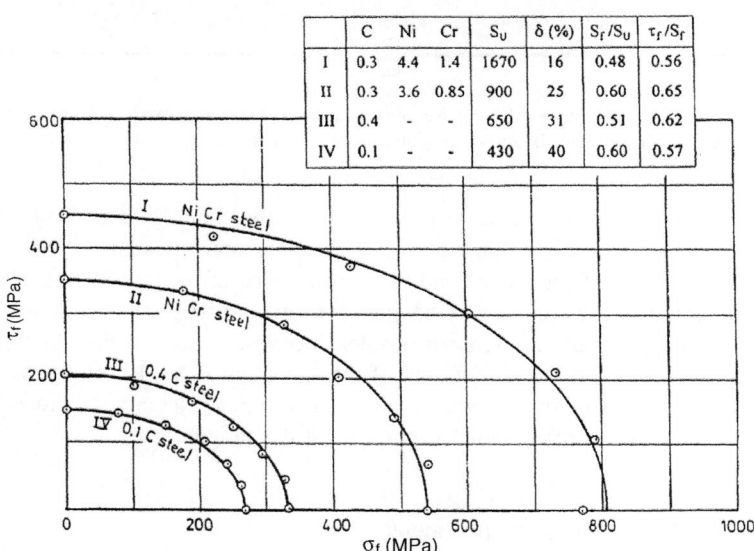

	C	Ni	Cr	S_u	δ (%)	S_f/S_u	τ_f/S_f
I	0.3	4.4	1.4	1670	16	0.48	0.56
II	0.3	3.6	0.85	900	25	0.60	0.65
III	0.4	-	-	650	31	0.51	0.62
IV	0.1	-	-	430	40	0.60	0.57

Fig. 14 The fatigue limit under proportional bending and torsion with zero mean stress. Results of Gough and Pollard [13]

(a)

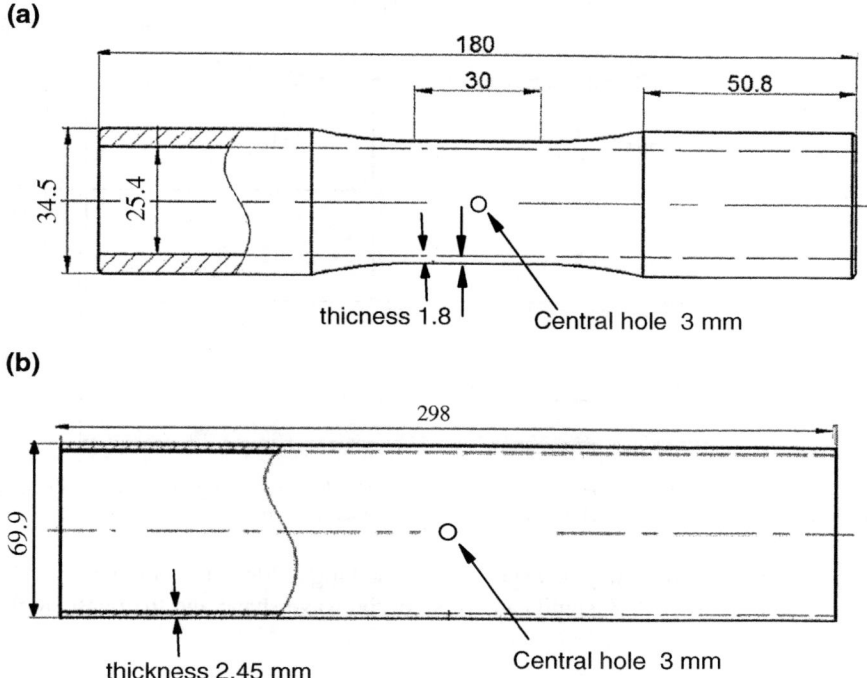

Fig. 15 Two different tubular aluminium alloy specimen adopted by Fatemi et al. [15].
a Machined tubular specimen. b Tubular specimen cut from an existing profile. Dimensions in
figures are presented in mm

designer this failure criterion is not realistic, because it is a comparison between
properties of small unnotched specimens and a real structural element.

Two different types of tubular specimens shown in Fig. 15 were adopted by
Fatemi et al. [15]. The upper specimens was machined from solid material whereas
the lower specimen was cut from extruded tube material. Crack growth was studied
under different biaxial load sequences.

Gough and Pollard suggested already in 1935 that a hollow specimen could also
be loaded by applying an internal pressure in the specimen. This is rarely done,
probably because it needs a high internal pressure to obtain a sufficiently large
circular hoop stress. However this type of biaxial loading was adopted by Dietmann
et al. [16]. The specimen is shown in Fig. 16. It has a relatively large diameter of the
circular section. The specimen was loaded by a longitudinal stress and a circumferential stress, which were also applied in phase and out-of-phase, and with different wave shape including phase angle shifts. The paper does not mention the
medium to apply the internal pressure but it is expected to be oil. Test results are
expressed in numbers of applied cycles until failure in S-N plots. The fatigue failure

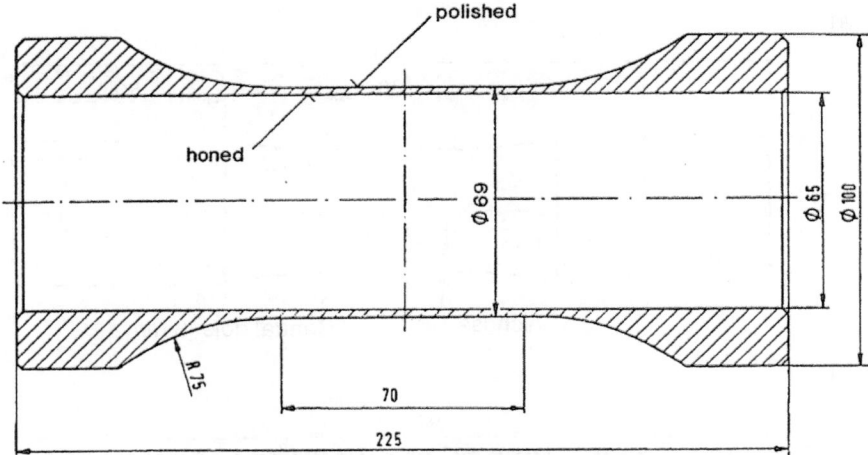

Fig. 16 Tubular specimen used by Dietmann et al. [16]. Relatively large diameter to obtain a significant hoop stress. Dimensions in figures are presented in mm

is not described, but it may be expected that the fatigue life is the number of cycles until leakage occurred. A significant effect of the wave shape and the phase angles was observed.

3.2.2 Notched Specimens

Solid notched specimens have been adopted in some investigations to observe a notch effect under biaxial loading. A specimen introduced by Sonsino and Pfohl [17] is shown in Fig. 17. The shoulder fillet represents a well defined geometrical notch for which the stress concentration factors K_t for bending and for torsion are known. Sonsino et al. have carried out various tests with proportional biaxial loads and also tests with a phase angle shift of 90° between the two sinusoidal cyclic

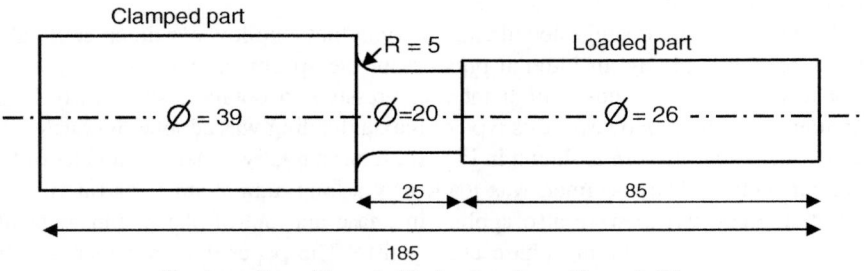

For bending: $K_{tb} = 1.49$. for torsion: $K_{tt} = 1.24$

Fig. 17 Notched specimen used by Sonsino and Pfohl [17]. Dimensions in figures are presented in mm

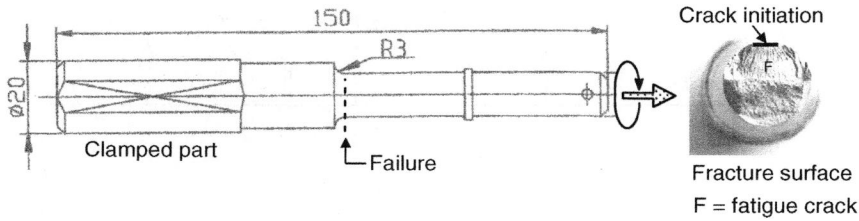

Fig. 18 Notched rotating beam specimen with constant torsion used by De Freitas et al. [9]. Dimensions in figures are presented in mm

loads. Tests at several load amplitudes were carried out to obtain *S-N* curves. In some test series the initially invisible growth of a small fatigue crack was recorded with a DC potential drop technique. The fatigue life until a small crack with a length of 1.0 mm was obtained and associated with a crack initiation life. The remaining life until failure is then the crack growth life. It may be noted here that the specimen of Sonsino can be used to explore the effect of several conditions associated with various practical fatigue conditions, such as notch root radius, size effect, effect of surface quality, etc.

An other notched specimen in shown in Fig. 18. It is a rotating beam specimens used by De Freitas et al. [9, 10] to study the effect of a superimposed torsion moment as discussed in Sect. 2.3.

3.2.3 Crack Growth Specimens

Research on fatigue crack growth on sheet specimens under biaxial loads has been stimulated by different arguments. Crack propagation with crack length measurements can easily be carried out. Furthermore the suggested validity of fracture mechanics procedures was also promoting various research programs. Moreover, the crack growth life in full-scale structural elements is important for damage tolerance of a structure in service.

Simple sheet specimens with oblique through cracks shown in Fig. 19 can be used to obtain basic information about fatigue crack growth under biaxial load sequences. The specimen with the central oblique crack in Fig. 19b appears to have an elementary character. A fully different specimen is the so-called ARCAN specimen shown in Fig. 20. It is also labelled as a butterfly specimen. The specimen can be mounted in the loading test rig under different angles in steps of 15°. For a through crack a relevant test result is the crack path. The specimen was adopted by Galyon et al. [18] to investigate small part through cracks which implies that the stress intensity varies along the crack front, a complicated condition. For through cracks it is generally suggested that fatigue cracks will growth perpendicular to the maximum principal tensile stress.

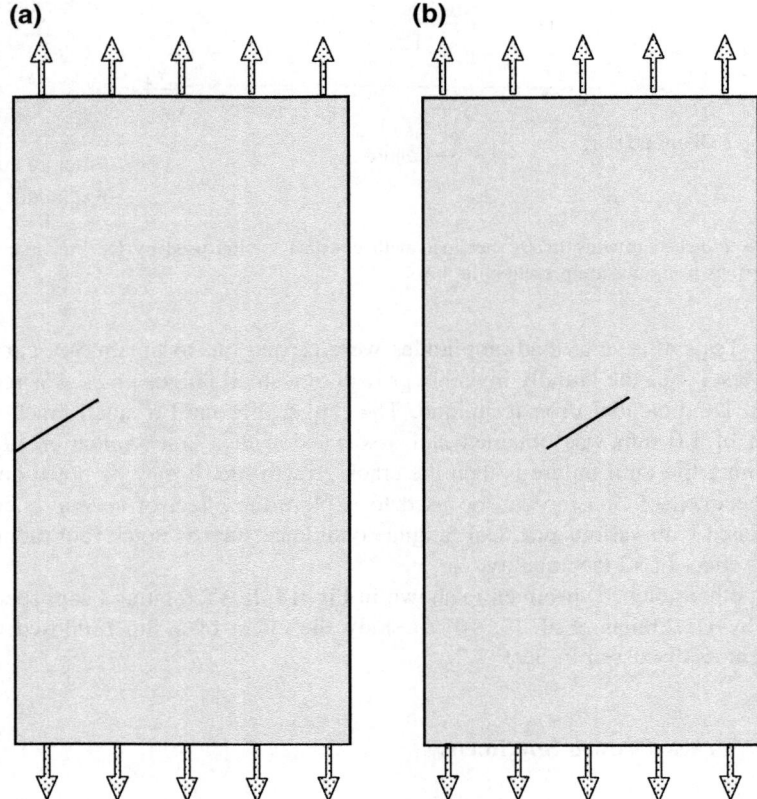

Fig. 19 Specimens with oblique cracks. **a** Oblique edge crack. **b** Oblique central crack

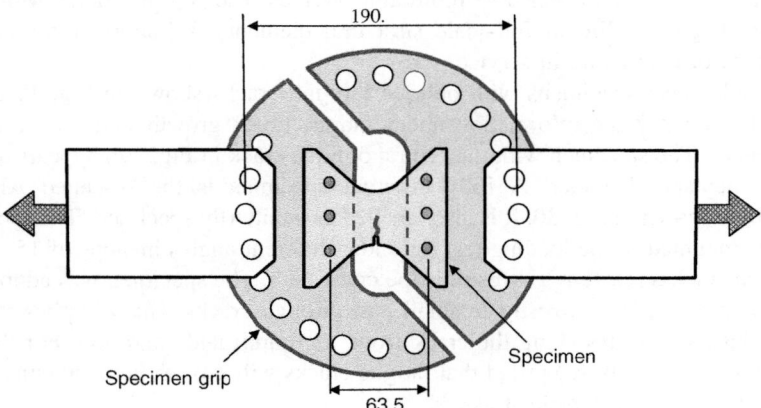

Fig. 20 ARCAN specimen in a test fixture for different loading direction used by Galyon et al. [18]. Dimensions in figures are presented in mm

It may be recalled here that fatigue cracks in specimens are starting as a small part through crack. Under biaxial loading it will imply that the stress intensity factor varies along the crack front. It is not physically obvious that predictions with fracture mechanics will then provide a rational solution.

3.2.4 Cruciform Specimen

The cruciform specimen appears to be a logical approach for research on fatigue under two biaxial tension loads. Unfortunately it offers a significant problem because a severe stress concentration occurs at the armpits where the two load flows are meeting. Large notch radii were adopted for the cruciform specimen used by Yuuki et al. [14], see Fig. 21. Tests were carried out in salt water and in air.

An entirely different type of a specimen was adopted by Dalle Donne and Döker [19], see Fig. 22. Slits were made all around the specimen edges which reduces the lateral stiffness of the edges to promote a more homogeneous load distribution on the central test section. Moreover, the thickness of the inner part of this section was reduced. The investigation was primarily focussed on the development of a specimen with a homogeneous stress distribution in the central section of the specimen. The specimen is expensive. Tests were associated with fracture toughness issues.

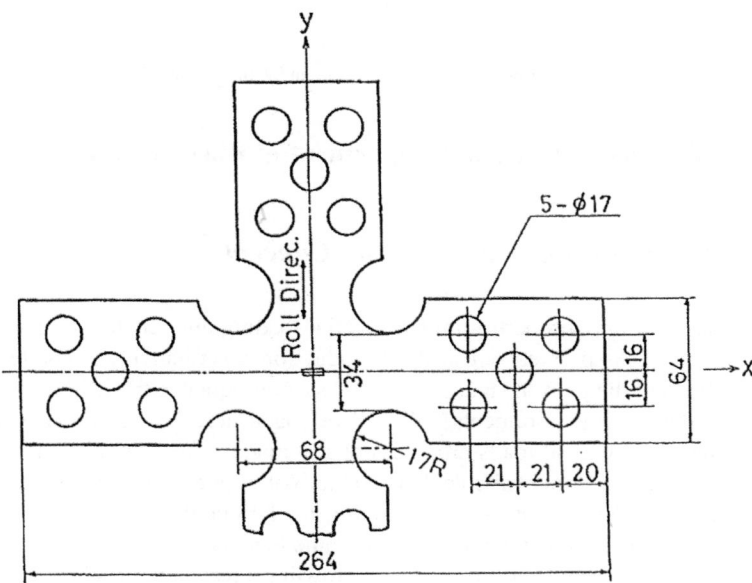

Fig. 21 Cruciform specimen with large radii of the armpits and a crack starter notch [14]. Dimensions in figures are presented in mm

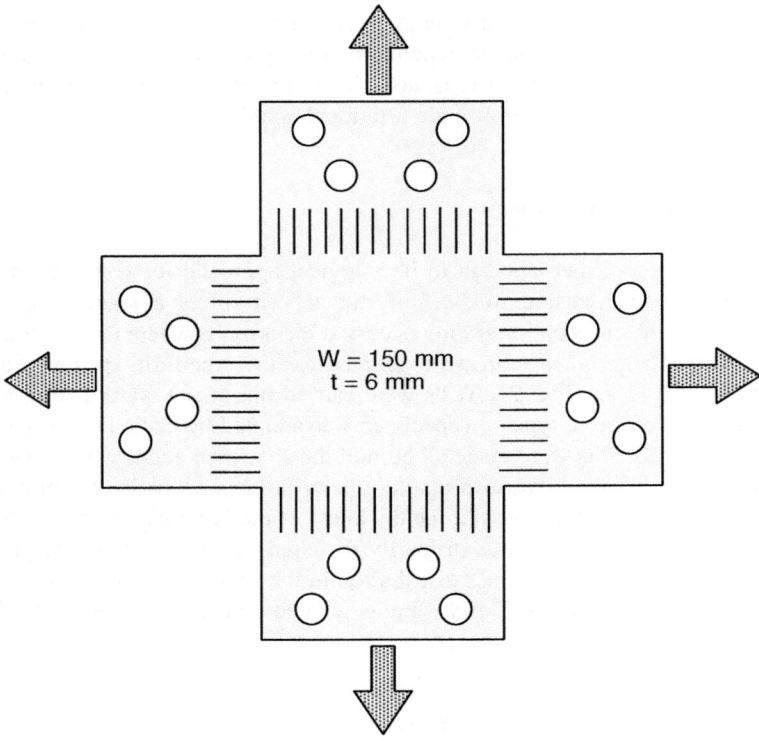

Fig. 22 Crucifix specimen proposed by Dalle Donne and Döker [19]

4 Predictions of Fatigue Properties for Biaxial Fatigue Loads

4.1 Predictions and the Similarity Concept

Predictions of fatigue properties under uniaxial loading are generally considered to be problematic. As pointed out in the introduction quantitatively accurate predictions are beyond the present state of the art. As a consequence this will also apply to biaxial loading. Some comments will still be made here because suggestions for predictions under biaxial loads are occasionally made in the literature. Predictions are an extrapolation from available data. Actually it implies that a similarity concept is used. Two similar systems are compared. If the fatigue damage of a load cycle in one system is known, it is assumed that the same load cycle in the other system will cause the same damage. For fatigue crack growth it means that a ΔK-cycle causing a crack length increment Δa in a specimen will cause the same Δa in an other specimen if the same ΔK-cycle is applied. This is a basic rule of fracture mechanics. The similarity approach for biaxial fatigue requires that the fatigue damage for the basic system can be described in physical terms. However, the only available information is associated with the τ/σ-path experience. It does not enable a physical description of the fatigue damage obtained. As a consequence, predictions based on a similarity concept are impossible.

4.2   Biaxial Fatigue of Full-Scale Structures

Designing of a structure against fatigue will remain a challenge for the design office in the industry. The structure to be delivered to the operator should not meet with fatigue problems in service. It is noteworthy that engineers in a design office do not trust theoretical predictions on fatigue properties. They are familiar with various aspects of the structure which can affect the fatigue performance. This is especially true for joints in a structure which frequently are the most fatigue critical items. But it can also apply to other fatigue sensitive details of the structure associated with production aspects and unexpected design errors. A carefully planned FE analysis can provide useful information for reducing stress concentrations. However, realistic information about the fatigue performance of a structure can not be obtained with calculations only. It requires realistic full-scale service-simulation fatigue experiments. Such tests are now generally accepted for aircraft, automotive vehicles, offshore structures, pressure vessels, wind mill installations, and investigations of various types of joints. Another practical approach is to run prototype vehicles under severe conditions in realistic operational circuits. It illustrates that the industry is prepared to invest much efforts to arrive at realistic information about the fatigue performance of structures in service. This can also be understood in view of safety and liability arguments. Within the present scope of the problem setting, the interest for theoretical predictions of fatigue properties of a structure is absent.

An interesting research investigation on a large full-scale structure was recently published by Sonsino [20]. It is summarized here because it covers the overall analysis of fatigue of a large welded structure. Fatigue tests were carried out on K-node elements of an offshore structure, see Fig. 23. The tests were performed with a service-simulation variable-amplitude load history based on an offshore load spectrum. A sea-water environment was applied which further increased the realistic nature of the experiments. A stress analysis of the structure with FE calculations was also made to reveal the most critical location for crack initiation in the welds, the so-called "hot spot". Strain gage measurements were made for the same purpose. In the experiments observations were made on crack initiation and subsequent crack growth. The crack initiation period was supposed to be finished after a crack with a crack depth of 1.0 mm under the material surface was obtained. The crack depth was measured with a DC potential drop technique. The end of the crack growth life was defined as the break through of the fatigue crack until the full wall thickness of the local structure. Such results can not be obtained by predictions.

Another practical problem was investigated by De Freitas and Da Fonte [9, 10]. Results were already mentioned in the discussion of Fig. 18. Power driving axles

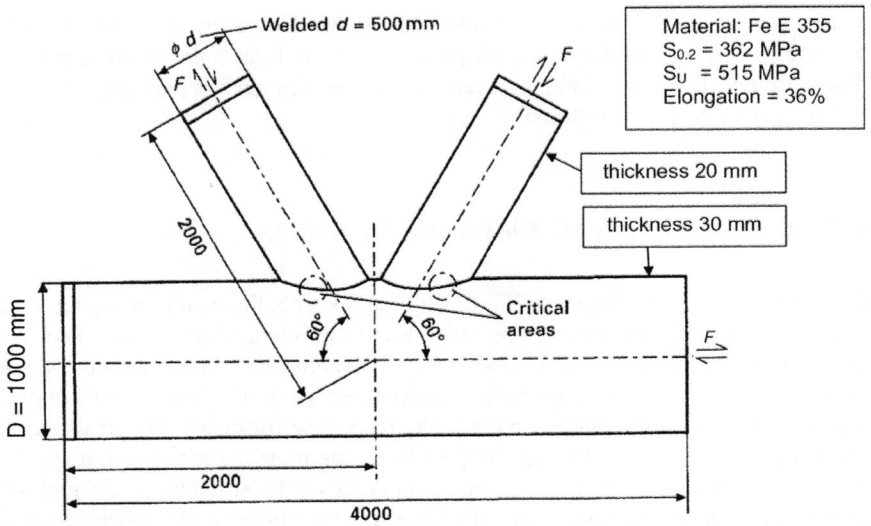

Fig. 23 Tubular welded K-node in an off-shore structure tested by Sonsino [20]

are present in various types of structures, for instance for transportation on railways, and for dredging machines. It also occurs in a driving axle assembled with a misalignment. A major load will be a steady torsion moment while many axles are also loaded in cyclic bending during rotation of the axle. Each rotation is a bending cycle which together with the steady torsion creates a biaxial load condition. Also in this case experiments are useful because predictions can not be based on crack growth data obtained under exactly similar loading conditions.

Some basic understanding about biaxial fatigue can also be helpful if a fatigue failure occurs in service. The failure analysis should include a careful examination of the fracture surface, macroscopically and microscopically. It can give useful information about the type of loading which has caused the failure. In general fatigue failures are easily recognized. It may also reveal whether it was due to load cycles in tension, bending or torsion. Indications of biaxial loading can also be obtained by a microscopic examination of the fatigue fracture surface, for instance if damage of crack surface interference is observed such as the sliding damage discussed in Sect. 2.3.

5   Summarizing Conclusions

The purpose of the present paper is to explore the significance of the extensive literature on biaxial fatigue of metallic materials and structures. Fatigue under variable-amplitude loading is not considered because fatigue damage accumulation under uniaxial variable-amplitude fatigue loads is not a really solved problem. As a consequence research papers about constant-amplitude loading are considered only. It still leaves a very large number of publications, also because the number of variables is large. Think of different specimens, different materials, and last but not least a large variety of biaxial load sequence adopted in research programs. Bolted, riveted and bonded joints do not allow a rational analysis of biaxial fatigue. They are not considered in the present paper. Also welded joints are not included in view of the large variety of these joints, although one case history is discussed in Sect. 4.2.

It should be understood that the present paper can not be a complete survey of all relevant publications in the literature. A selection of papers has been made to be studied in more detail, mainly more recent publications, but some older ones as well. The selection implies that equally informative and valuable papers are not included in the list of references. The author hopes to be excused for the selection made. It is not a denial of many useful and interesting papers.

The present paper starts with a discussion of physical aspects of the fatigue phenomenon in metallic materials because physical understanding is essential to justify fatigue concepts. It is followed with a description of biaxial load sequences in terms of τ/σ-paths. Various types of specimens adopted for biaxial fatigue tests are described. Results of biaxial fatigue tests are discussed and noteworthy observation are recapitulated. It has led to the following conclusions.

1. The fatigue life of a specimen or a structure covers three successive periods: the crack nucleation period, the initial micro crack period and the macro crack propagation period. Crack nucleation starts with a microscopically small part through crack. The initially growth in the second phase can still be somewhat irregular. In the third phase starting at a crack size in the order of 1 mm a more continuous crack growth occurs until failure. In all three phases cyclic slip occurs which is converted into crack nucleation and subsequent crack growth. The decohesion mechanism is a matter of speculation, but it should be expected that the fatigue mechanism will be affected by the biaxial stress conditions.

2. Cyclic slip can lead to cyclic strain hardening and cyclic strain softening depending on the material structural characteristics. Anisotropy of the material can have a large effect on the biaxial fatigue behaviour of a material. Generalization of results is questionable.

3. A significant effect of a phase angle shift between the two biaxial loads has been noted in several research programs. It appears to be logical that the conversion of cyclic slip into crack nucleation and crack extension will depend on the simultaneous stress perpendicular to the slip plane. It explains why the results are different for proportional and non-proportional biaxial loads.

4. An outstanding observation is the so-called crack surface interference (CSI). Crack growth can occur in different modes. As an example crack growth in mode III under cyclic torsion loads will be affected by a stress perpendicular to the growing fatigue crack. It will open or close the crack depending on the question whether it is a tensile stress or a compression stress. In the latter case

cyclic sliding between the two fatigue fracture surfaces will occur which will reduce the crack growth rate. The crack surface interference can also occur for open cracks if an undulated fatigue crack topology is present, such as a factory roof fracture surface caused by cyclic torsion. Damage of cyclic sliding crack movements can be visible in the SEM.

5. Prediction of biaxial fatigue properties require a similarity concept. It implies that biaxial fatigue damage should be explicitly defined in physical terms. In spite of knowing the τ/σ-path of every cycle and of some understanding of critical slip planes, the accumulation of fatigue damage can not be defined in physical terms. As a consequence, predictions can not be trustworthy.

6. Biaxial stress condition occur in all engineering structures. Research in the future can still shed more light on biaxial fatigue phenomenon. It is desirable that low magnification pictures of the fatigue failure and SEM pictures of the fatigue fracture surfaces are presented in research papers in order to see what happens. Furthermore, it is also desirable to focus more attention to engineering design variables, such as notch effects and characteristic material conditions.

7. Engineers in a design office of the industry do not trust predictions. Designing against fatigue is optimized by FE analysis. Questions about fatigue are solved by realistic service-simulation fatigue tests. The apparent gap between the research approach and the engineering perception should be reconsidered.

References

1. Schijve J (2009) Fatigue of structures and materials, 2nd edn. Springer, Berlin
2. Schijve J (2003) Fatigue of structures and materials in the 20th century and the state of the art. Int J Fatigue 25:679–702
3. Marquis G (2007) Current trends in multiaxial fatigue research and assessments. In: 8th international conference multiaxial fatigue and fracture, Sheffield
4. Shamsaei N, Fatemi A (2014) Small fatigue crack growth under multiaxial stresses. Int J Fatigue 58:126–135
5. Schijve J (2009) Fatigue as a phenomenon in the material. In: Fatigue of structures and materials, 2nd edn. Springer, Berlin
6. Tanaka K (2014) Crack initiation and propagation in torsional fatigue of circumferential notched steel bars. Int J Fatigue 58:114–125
7. Tschegg EK (1983) Sliding mode crack closure and mode II fatigue crack growth in mild steel. Acta Metall 31:1323–1330
8. Yu X, Abel A (2000) Crack growth interference under cyclic mode I and steady mode II loading. Part I: experimental study. Eng Fract Mech 66:503–518
9. De Freitas M, Reis L, Da Fonte M, Li B (2011) Effect of steady torsion on fatigue crack initiation and propagation under rotating bending. Multiaxial fatigue and mixed-mode cracking. Eng. Fract Mech 78:826–835
10. Da Fonte M, Reis L, De Freitas M (2014) The effect of steady torsion on fatigue crack growth under rotating bending loading on aluminium alloy 7075-T6. Frattura ed Integrita Strutturale 30:360–368
11. Socie DF, Marquis G (2000) Multiaxial fatigue. Society of Automotive Engineers, Warrendale, USA

12. Fatemi A (2001) Multiaxial stresses, chapter 10. In: Stephens RI, Fatemi A, Stephens RR, Fuchs HO (eds) Metal fatigue in engineering. Wiley, New York
13. Gough HJ, Pollard HN (1935) The strength of metals under combined alternating stresses. In: Proceedings of the Institution of Mechanical Engineers, vol. 131, London
14. Yuuki R, Murakami E, Kitagawa H (1989) Corrosion fatigue crack growth under biaxial stresses. EGF 3:285–300 (Mech. Eng. Publication, London)
15. Fatemi A, Gates N, Socie DF, Phan N (2014) Fatigue crack growth behaviour of tubular aluminium specimens with a circular hole under axial and torsion loadings. Eng Fract Mech 123:137–147
16. Dietmann H, Bhongbhibhat T, Schmidt A (1991) Multiaxial fatigue behaviour of steels under in-phase and out-of-phase loading, including different wave forms and frequencies. In: Kussmaul K, McDiarmid D, Socie D (eds) Fatigue under biaxial and multiaxial loading. ESIS 10. Mechanical Engineering Publications, London, pp 465–475
17. Sonsino CM, Pfohl R (1991) Multiaxial fatigue of welded shaft-flange connections of stirrers under random non-proportional torsion and bending. In: Kussmaul K, McDiarmid D, Socie D (eds) ESIS 10. Mechanical Engineering Publications, London, pp 449–464
18. Galyon SE, Arunachalam R, Greer J, Hammond M, Fawaz SA (2009) Three dimensional crack growth prediction. Bridging the gap between theory and operational practice. ICAF Symposium 2009. Springer, Berlin, pp 1035–1068
19. Dalle Donne D, Döker H (1994) Biaxial load effects on plane stress J-Δa- and δ_5-Δa-curves. ECF 10, Structural integrity, vol 11, EMAS
20. Sonsino CM (2012) Comparison of different local design concepts for the structural durability assessment of welded offshore K-nodes. Int J Fatigue 34:27–34